"Life is 50% aptitude and 50% attitude".
Yusnier Viera.

Basic Course of Mental Arithmetic

Yusnier Viera

www.spicymath.com

Contents

1 Simple multiplication and division.

We are all capable of learning how to do mental calculations quickly and easily. The problem is that most people don't know the correct ways to do them and they don't have the confidence in their mental capabilities to even attempt them.

For the most part, most of us have difficulties doing most, if not all, mental calculation. Often times we choose not to do them and instead use a calculator, computer, or smartphone. Even when we do commit to a mental calculation, we still tend to spend a lot of time doing it and/or we do it with very little certainty of whether our answer is correct or not.

In this book, the reader will learn the most efficient ways to mentally calculate multiplication, division, and square root problems. Not only that, but the reader will develop a better sense of their mental capabilities while also developing a healthy confidence in them. By the end of this book you should able to quickly perform numerical operations mentally with ease.

1.1 Multiplication by powers of 10.

When multiplying large numbers or decimals, there are certain cases where we can solve the problem by doing some simple arithmetic. Because we can reduce the problem to a much simpler task, the answer can be obtained much faster.

Multiplying a number by 10 is easy to do. Let's look at some examples:

Ex. $26 \times 10 = 260$ (Multiplying by 10 is equivalent to adding a zero to the other number).

Multiplying by 1000 is just as easy:
Ex. $43 \times 1000 = 43000$ (Multiplying by 1000 is equivalent to adding three zeros to the other number).

In general, when we multiply by any number that is a one followed by zeros, we get the result by adding the number of zeros to the other number.

Ex. $3567 \times 10000 = 35670000$

In the case of a decimal number:
Ex. $135.478 \times 10000 = 1354780$
Because the number has three decimal places, we add just one zero. Note that adding a zero is equivalent to moving the decimal place to the right one spot.

It's pretty remarkable how simple it is to multiply by a power of 10. Have you ever wondered why?

Well, it's because the system we use to represent numbers is the **decimal** system, which is composed by 10 digits: 0, 1, 2, 3, 4, 5, 6, 7, 8, 9. This makes multiplying by any power of 10 (10, 100, 1000) very easy.

1.2 Division by powers of 10.

Similarly, division by powers of 10 is easy.

Examples:

$\frac{36}{10} = 3.6$ (equivalent to moving the decimal one place to the left).

$\frac{23}{1000} = 0.023$ (equivalent to moving the decimal three places to the left).

Dividing by 10 is easy because we use a decimal system. There are other quick mental calculations we can do by taking advantage of the decimal system.

1.3 How to multiply doing divisions?

1.3.1 How to multiply by 5?

Rule of multiplication by 5:

$$5N = \frac{N}{2} \times 10 \tag{1}$$

Multiplying by 5 is the same as dividing by 2 and then multiplying by 10 (or vice versa), because $\frac{10}{2} = 5$. Applying this rule makes multiplication by 5 simpler.

Example 1:

$14 \times 5 = \frac{14}{2} \times 10 = 7 \times 10 = 70$

The result of the multiplication of 14×5 is equal to 70.

Example 2:

$222 \times 5 = \frac{222}{2} \times 10 = 111 \times 10 = 1110$

In summary, when we multiply a number by 5 we divide the number by 2 and then we add a zero to get the final answer. Now the question becomes, how easily can we divide by 2 so that our multiplication by 5 is quick.

1.3.2 How to divide by 2?

Example 1:

$\frac{2468}{2} = \ldots$

The method is simple. We take the first digit of the left and divide by 2:

$\frac{2468}{2} = 1 \ldots$

Then we take the second digit of the left and divide by 2:

$$\frac{24\cancel{6}8}{2} = 12\ldots$$

And so on until the last digit:

$$\frac{2468}{2} = 1234$$

The result is 1234.

What happens if the digit we need to divide by 2 is odd? In that case we only need to do a small adjustment.

Example 2:

$$\frac{652}{2} = \ldots$$

We take the first digit of the left, the digit 6, and divide by 2:

$$\frac{652}{2} = 3\ldots$$

Then we take the second digit of the left and we notice it is an odd number. In consequence:

$\frac{6\cancel{5}\cancel{2}}{2} = 3\ldots$ We take away 1 from the second digit, which is 5, and we add 10 to the next digit (the 5 becomes a 4 and the next digit, which is 2, becomes 12).

And so the procedure continues:

$$\frac{64(12)}{2} = 3\ldots$$

$\frac{64(12)}{2} = 32\ldots$ then we find the half of the next digit that would be 12, the half of 12 is 6.

$$\frac{64(12)}{2} = 326$$

Therefore, the final result would be:

$$\mathbf{\frac{652}{2} = 326}$$

Example 3:

$$\frac{78934}{2} = \ldots$$

The first digit 7 is odd, then:

$\frac{\cancel{7}8934}{2}$ Take away 1 from 7 and add 10 to the next digit.

$$\frac{6(18)934}{2} = 3\ldots$$

$\frac{6(18)934}{2} = 39\ldots$ The half of 18 is 9.

$\frac{6(18)\cancel{9}34}{2} = 39\ldots$ The digit 9 is odd, then
$\frac{6(18)8(13)4}{2} = 394\ldots$

$\frac{6(18)8\cancel{(13)}4}{2} = 394\ldots$ The number 13 is odd, then

$\frac{6(18)8(12)(14)}{2} = 3946\ldots$

$\frac{6(18)8(12)(14)}{2} = 39467$

Therefore:

$$\frac{78934}{2} = \mathbf{39467}$$

Example 4:

$\frac{62347}{2}$

$\frac{62347}{2} = 3\ldots$

$\frac{62347}{2} = 31\ldots$

$\frac{62\cancel{3}47}{2} = 31\ldots$ The digit 3 is odd, then

$\frac{622(14)7}{2} = 311\ldots$

$\frac{622(14)7}{2} = 3117\ldots$

$\frac{622(14)\cancel{7}}{2} = 3117\ldots$ The digit 7 is odd and it is the last one, then

$\frac{622(14)6,(10)}{2} = 31173,\ldots$ Since there is no other digit, we assume the next digit is 0 because $62347=62347.0$

$\frac{622(14)6,(10)}{2} = 31173.5$

Therefore:

$$\frac{62347}{2} = \mathbf{31173.5}$$

1.3.3 Back to the multiplication by 5.

Now that we know how to divide by 2, we are ready to perform any multiplication by 5. Remember the rule for multiplying by 5:

$$5N = \frac{N}{2} \times 10 \qquad (2)$$

Example:

365×5 — First we divide by 2 and then we multiply by 10.

$$\frac{\cancel{3}65}{2} = \ldots$$

$$\frac{2(16)5}{2} = 1\ldots$$

$$\frac{2(16)5}{2} = 18\ldots$$

$$\frac{2(16)\cancel{5}}{2} = 18\ldots$$

$$\frac{2(16)4,(10)}{2} = 182,\ldots$$

$$\frac{2(16)4,(10)}{2} = 182.5$$

Then we multiply by 10:
$182.5 \times 10 = 1825$ Therefore,

$365 \times 5 = 1825$

Let's consider some other examples, now with decimal numbers.

Example 1:

$73.266 \times 5 = \ldots$

$$\frac{\cancel{7}3.266}{2} = \ldots$$

$$\frac{6(13).266}{2} = 3\ldots$$

$$\frac{6(\cancel{13}).266}{2} = 3\ldots$$

$$\frac{6(12),(12)66}{2} = 36,\ldots$$

$$\frac{6(12),(12)66}{2} = 36,6\ldots$$

$$\frac{6(12),(12)66}{2} = 36.63\ldots$$

$$\frac{6(12).(12)66}{2} = 36.633$$

$36.633 \times 10 = 366.33$ Therefore,

$$\mathbf{73.266 \times 5 = 366.33}$$

Example 2:

$324.6 \times 5 = \frac{324.6}{2} \times 10 = 162.3 \times 10 = 1623$

Example 3:

$24.86 \times 5 = \frac{24.86}{2} \times 10 = 12.43 \times 10 = 124.3$

Example 4:

$0.0082 \times 5 = \frac{0.0082}{2} \times 10 = 0.0041 \times 10 = 0.041$

In the examples we just saw, the first thing we do is divide by 2 and then move the decimal one place to the right of the number obtained. This is the same as saying we multiply by 10 the number obtained in the previous step.

Here are some suggested exercises that can be done individually, applying what you learned.

Exercises:

a) $825 \times 5 =$
b) $42362 \times 5 =$
c) $3242446 \times 5 =$
d) $102 \times 5 =$
e) $1004 \times 5 =$
f) $305.4 \times 5 =$
g) $1084.24 \times 5 =$
h) $0.42 \times 5 =$
i) $14.87 \times 5 =$
j) $14.846 \times 5 =$

1.3.4 How to multiply by 25?

When multiplying any number by 25, we can get the same result if we divide that number by 4 and then multiply by 100. Note that $\frac{100}{4} = 25$. Here is the rule to follow when multiplying by 25:

$$25N = \frac{N}{4} \times 100 \qquad (3)$$

Examples:

$48 \times 25 = \frac{48}{4} \times 100 = 12 \times 100 = 1200$

$12 \times 25 = \frac{12}{4} \times 100 = 3 \times 100 = 300$

$840 \times 25 = \frac{840}{4} \times 100 = 210 \times 100 = 21000$

In summary, if we want to multiply a number by 25, we first divide it by 4 and then add two zeros to the number.

$32 \times 25 = \ldots$

$\frac{32}{4} = 8$ — Divide by 4.

$8 \times 100 = 800$ — Add two zeros.

Here are some other examples with decimal numbers.

Example 1:

$48.04 \times 25 = \frac{48.04}{4} \times 100 = 12.01 \times 100 = 1201$

Example 2:

$32.8 \times 25 = \frac{32.8}{4} \times 100 = 8.2 \times 100 = 820$

Example 3:

$164.24 \times 25 = \frac{164.24}{4} \times 100 = 41.06 \times 100 = 4106$

Example 4:

$0.00128 \times 25 = \frac{0.00128}{4} \times 100 = 0.00032 \times 100 = 0.032$

In practice, mentally we divide by 4 and then multiply by 100. In other words, we add two zeros if the number is an integer, or we move the decimal point two places to the right if it is a decimal number.

Now try some exercises on your own while applying what we just learned.

Multiply by 25 each of the following numbers:

a) 16
b) 52
c) 124
d) 400
e) 24.48
f) 12.24

1.3.5 How to multiply by 12.5?

Multiplying by 12.5 is the same as dividing by 8 and then multiplying by 100.

$$12.5N = \frac{N}{8} \times 100 \tag{4}$$

Example 1:

$56 \times 12.5 = \frac{56}{8} \times 100 = 7 \times 100 = 700$

Example 2:

$168 \times 12.5 = \frac{168}{8} \times 100 = 21 \times 100 = 2100$

Now let's do it faster.

Example 3:

$48 \times 12.5 = 6 \times 100 = 600$

Example 4:

$3288 \times 12.5 = 411 \times 100 = 41100$

Example 5:

$6448 \times 12.5 = 806 \times 100 = 80600$

Bear in mind the following when we are dividing by 8:

$\frac{64}{8} = 8$ and when you continue dividing, the number 4 is not divisible by 8 in the natural set of numbers, so put a zero in the quotient for continue dividing and it would be 48 divided by 8.

As we have seen in practice, to multiply a number by 12.5 first divide the number by 8 and then move the decimal two places to the right.

$16 \times 12.5 = \ldots$

$\frac{16}{8} = 2$ — Divide the number by 8.

$2 \times 100 = 200$ — Add two zeros.

Let's see what happens when we multiply 12.5 by decimal numbers.

Example 1:

$32.88 \times 12.5 = \frac{32.88}{8} \times 100 = 4.11 \times 100 = 411$

Example 2:

$720.8 \times 12.5 = \frac{720.8}{8} \times 100 = 90.1 \times 100 = 9010$

As seen above, what we do is divide by 8 and then move the decimal point two places to the right.

Example 3:

$24.72 \times 12.5 =$

$\frac{24.72}{8} = 3.09$ — Divide the number by 8.

$3.09 \times 100 = 309$ — When multiplying by 100, we move the decimal point two places to the right.

Try the following exercises.

Multiply by 12.5 the following numbers:

a) 24
b) 48
c) 32
d) 72
e) 3248
f) 7288
g) 16.96
h) 248.4

It's very important to try all the exercises because it will help you acquire and develop skills for solving them mentally.

1.4 How to multiply by other numbers?

1.4.1 Multiplication by 11.

Multiplying a two-digit number by 11 is simple. Here's the process:

$35 \times 11 = 3\ 5$ — write the number 35 leaving a space in the middle.

To calculate the middle digit we add the digits on the outsides. As we have that $3 + 5 = 8$ then,

$35 \times 11 = 3(3{+}5)5$

$35 \times 11 = 385$

Simple, right? Let's try a few examples:

$42 \times 11 = 4(4{+}2)2$

$42 \times 11 = 462$

Let's do it directly in one step:

$54 \times 11 = 594$

$17 \times 11 = 187$

But... What happens when the sum of the digits of the sides results in a two-digit number? Consider the following example:

$39 \times 11 = 3(3{+}9)9$

$39 \times 11 = 3(12)9$

$39 \times 11 = 3(12)9$ — The 12 becomes 2 and add 1 to the previous digit.

$39 \times 11 = 429$

Here is another example:

$85 \times 11 = 8(13)5$

$85 \times 11 = 935$

Let's do it directly in one step:

$68 \times 11 = 748$

$77 \times 11 = 847$

$95 \times 11 = 1045$

Try the following exercises:

a) $26 \times 11 =$
b) $45 \times 11 =$
c) $83 \times 11 =$
d) $64 \times 11 =$
e) $78 \times 11 =$
f) $96 \times 11 =$

Perhaps you're wondering... How can I multiply any number by 11? Here's how.

Example 1:

$12345 \times 11 = \ldots$

$12345 \times 11 = \ldots 5$ We start by writing the first digit on the right.

$12345 \times 11 = \ldots(4+5)5$ Then we begin to add each new pair of digits.

$12345 \times 11 = \ldots 95$

$12345 \times 11 = \ldots(3+4)95$

13

$12345 \times 11 = \ldots 795$

$12345 \times 11 = \ldots(2{+}3)795$

$12345 \times 11 = \ldots 5795$

$12345 \times 11 = \ldots(1{+}2)5795$

$12345 \times 11 = \ldots 35795$

$12345 \times 11 = 135795$ Finally, we write the digit on the left.

Therefore:

$12345 \times 11 = 135795$

But... What if the sum gives me a two-digit number?

Example 2:

$267 \times 11 = \ldots$

$267 \times 11 = \ldots 7$

$267 \times 11 = \ldots(6{+}7)7$ Now the sum is a two-digit number, because 6+7=13.

$267 \times 11 = \ldots 37$ We write only the 3 and mentally we carry 1 to the next step.

$267 \times 11 = \ldots(2{+}6{+}1)37$ Add 2+6 and also add the 1 of the previous step.

$267 \times 11 = \ldots 937$ Remember that we have not finished yet, we are missing the digit on the left.

$267 \times 11 = 2937$ Finally, write the digit on the left.

Therefore:

$267 \times 11 = 2937$

Example 3:

$7483904 \times 11 = \ldots$

$7483904 \times 11 = \ldots 4$

$7483904 \times 11 = \ldots (0+4)4$

$7483904 \times 11 = \ldots 44$

$7483904 \times 11 = \ldots (9+0)44$

$7483904 \times 11 = \ldots 944$

$7483904 \times 11 = \ldots (3+9)944$ 3+9=12, we write the 2 and we carry 1 to the next step.

$7483904 \times 11 = \ldots (8+3+1)2944$ 8+3+1=12, we write the 2 and we carry 1 to the next step.

$7483904 \times 11 = \ldots (4+8+1)22944$ 4+8+1=13, we write the 3 and we carry 1 to the next step.

$7483904 \times 11 = \ldots (7+4+1)322944$ 7+4+1=12, we write the 2 and we carry 1 to the next step.

$7483904 \times 11 = (7+1)2322944$ — Write the leftmost digit and add 1 from the previous step.

$7483904 \times 11 = 82322944$

Therefore:

$7483904 \times 11 = 82322944$

For in-depth knowledge of the algorithm, we encourage the reader to solve the next exercises:

g) $429 \times 11 =$
h) $62937 \times 11 =$
i) $84703 \times 11 =$
j) $3904168265 \times 11 =$

1.4.2 Multiplication by multiples of 11.

Now that we know how to multiply any number by 11, let's find a general method for multiplying by any multiple of 11.

The first positive multiples of 11 are:
11; 22; 33; 44; 55; 66; 77; 88; 99

By now we know how to multiply by 11, but... How do we multiply by 22, 33, 44, etc?

The algorithm that we will use is similar to the algorithm for multipliying by 11 explained above. Suppose we want to multiply by 33. The algorithm will be exactly the same as that used above, with the exception that at each step is necessary to multiply by 3, because $33=3\times 11$

Example 1:

$342 \times 33 = \ldots$

$34\cancel{2} \times 33 = \ldots6$ — Write the rightmost digit multiplied by 3.

$3\cancel{42} \times 33 = \ldots86$ $(4+2)\times3 = 18$, write 8 and carry 1 to the next step.

$\cancel{342} \times 33 = \ldots286$ $(3+4)\times3 + 1 = 22$, write 2 and carry 2 to the next step.

$\cancel{342} \times 33 = 11286$ $3\times3 + 2 = 11$, which is the last step.

Therefore:

$342 \times 33 = 11286$

Example 2:

$643 \times 88 =$

$64\cancel{3} \times 88 = \ldots4$ $8\times3=24$, write 4 and carry 2 to the next step.

$6\cancel{43} \times 88 = \ldots84$ $(4+3)\times8 + 2 = 58$, write 8 and carry 5 to the next step.

$\cancel{643} \times 88 = \ldots584$ $(6+4)\times8 + 5 = 85$, write 5 and carry 8 to the next step.

$\cancel{643} \times 88 = 56584$ $6\times8 + 8 = 56$.

Therefore:

$643 \times 88 = 56584$

Example 3:

$35048 \times 44 = \ldots$

$3504\cancel{8} \times 44 = \ldots2$ $8\times4=32$, write 2 and carry 3 to the next step.

$350\cancel{48} \times 44 = \ldots12$ $(4+8)\times4 + 3 = 51$, write 1 and carry 5 to the next step.

$35\cancel{048} \times 44 = \ldots112$ $(4+0)\times4 + 5 = 21$, write 1 and carry 2 to the next step.

16

3$5048$ × 44 = ...2112 (5+0)×4 + 2 = 2$2$, write 2 and carry 2 to the next step.

3$5048$ × 44 = ...42112 (3+5)×4 + 2 = 3$4$, write 4 and carry 3 to the next step.

35048 × 44 = 1542112 3×4 + 3 = 15.

Therefore:

35048 × 44 = 1542112

We propose the following exercises for individual work:

a) 718 × 22 =
b) 6603 × 55 =
c) 38247 × 77 =
d) 5082643 × 99 =

1.4.3   Multiplication by multiples of 9.

Multiplying by 9; 18; 27; 36; 45; 54; 63; 72; 81 is equivalent to multiply by:

$(10 - 1), (20 - 2), (30 - 3), (40 - 4), (50 - 5), (60 - 6), (70 - 7), (80 - 8), (90 - 9)$, respectively.

Example 1:

123 × 27 = 123 × (30 − 3) — If we apply the distributive property, we obtain:
= 123 × 30 − 123 × 3
= (123 × 3) × 10 − 123 × 3
= 369 × 10 − 369
= 3690 − 369
= 3321

Example 2:
— Now let's solve the example using a more direct way.

28 × 36 = 28 × (40 − 4)
= 28 × 40 − 28 × 4
= 1120 − 112
= 1008

Example 3:

$32 \times 45 = 32 \times (50 - 5)$
$= 1600 - 160$ — Here we apply the distributive property and we calculate the result mentally.
$= 1440$

Example 4:

$14 \times 18 = 14 \times (20 - 2)$
$= 280 - 28$
$= 252$

Example 5:

$32 \times 81 = 32 \times (90 - 9)$
$= 2880 - 288$
$= 2592$

Example 6:

$35 \times 54 = 35 \times (60 - 6)$
$= 2100 - 210$
$= 1890$

We can see with all these examples that while applying the distributive property, you need only perform the multiplication of the second term, since the first term is equal to the second term with a zero added, which is equivalent to multiplying by 10. The multiplication result is equal to the difference between these two numbers.

It is important to remember that multiplying by:

9 is equal to multiplying by (10 – 1)
18 is equal to multiplying by (20 – 2)
27 is equal to multiplying by (30 – 3)
36 is equal to multiplying by (40 – 4)
45 is equal to multiplying by (50 – 5)
54 is equal to multiplying by (60 – 6)
63 is equal to multiplying by (70 – 7)
72 is equal to multiplying by (80 – 8)
81 is equal to multiplying by (90 – 9)

You must remember these equivalences so you can apply the calculation procedures previously mentioned.

Example 1:

$23 \times 18 = 460 - 46$ — Because $18 = 20 - 2$
$= 414$

Example 2:

$330 \times 27 = 9900 - 990$ — Because $27 = 30 - 3$
$= 8910$

Try the exercises below.

Multiply:

a) $18 \times 45 =$
b) $21 \times 54 =$
c) $15 \times 63 =$
d) $34 \times 27 =$
e) $92 \times 36 =$
f) $43 \times 45 =$
g) $14 \times 54 =$
h) $23 \times 63 =$
i) $54 \times 72 =$
j) $35 \times 81 =$
k) $320 \times 27 =$
l) $121 \times 45 =$
m) $184 \times 45 =$
n) $143 \times 36 =$

NOW YOU ARE ABLE TO SOLVE NUMEROUS MULTIPLICATION PROBLEMS MENTALLY.

Multiply and apply what was learned:

1) $385 \times 5 =$
2) $436 \times 5 =$
3) $38.46 \times 5 =$
4) $602.8 \times 5 =$
5) $488.4 \times 5 =$
6) $2084.64 \times 5 =$
7) $720 \times 25 =$
8) $64 \times 25 =$
9) $16.88 \times 25 =$
10) $32 \times 12.5 =$
11) $16 \times 12.5 =$
12) $111 \times 27 =$
13) $26 \times 36 =$

14) $44 \times 63 =$
15) $14 \times 18 =$
16) $143 \times 45 =$
17) $26 \times 54 =$
18) $32 \times 72 =$
19) $48 \times 81 =$
20) $184 \times 27 =$
21) $143 \times 18 =$

With the knowledge you have acquired you can now mentally multiply any number. With practice you'll be able to do it even faster.

1.5 How to divide doing multiplications?

In this chapter, you will learn how to divide mentally quickly and accurately.

Here we will study some algorithms that will allow you avoid ever having to use a calculator. Mastering the algorithms will help you perform the calculations with excellent speed while also exercising your memory. The only requirement is to know the basic calculations that you've already learned in your first years of studies.

The goal is for you to learn how to calculate faster and safer.

1.5.1 How to divide by 5?

Dividing by 5 is the same as multiplying by 2 and then dividing by 10.

$$\frac{N}{5} = \frac{2N}{10} \tag{5}$$

For example. Divide:

$\frac{14}{5} = (14 \times 2) : 10 = 28 : 10 = 2.8$

Here are some example to analyze the algorithm:

Example 1:

$\frac{111}{5} = \ldots$

$111 \times 2 = 222$ — First multiply by 2.

$\frac{222}{10} = 22.2$ — When dividing by 10 you move the decimal one place to the left.

Then the final result is 22.2.

Example 2:

$\frac{143210}{5} = \ldots$

$143210 \times 2 = 286420$

$\frac{286420}{10} = 28642$

The result is 28642.

Example 3:
Try to solve it mentally.

$\frac{234}{5} = \ldots$

468 — Multiply by 2.

46.8 — Divide the number by 10.

The result is 46.8.

Using this same procedure, we will try division by 5, but with decimal numbers.

Remember, first multiply by 2 and then divide by 10.

Example 1:

$\frac{4.8}{5} = \ldots$

$4.8 \times 2 = 9.6$ — Multiply the number by 2.

$\frac{9.6}{10} = 0.96$ — When dividing by 10 you move the decimal point one place to the left.

The result is 0.96.

Example 2:

$\frac{81.3}{5} = (81.3 \times 2) : 10 = 162.6 : 10 = 16.26$

Example 3:

$$\frac{34.3}{5} = (34.3 \times 2) : 10 = 68.6 : 10 = 6.86$$

It seems that division by 5 is relatively easy. Essentially, it consists of multiplying by 2 and then divide by 10. As we have seen before, the division by 10 is natural because of the fact that we use a decimal system for our numbers. But ... Is there a simple algorithm to multiply by 2, so that you can speed up the calculation of dividing by 5? Fortunately the answer is yes.

1.5.2 How to multiply by 2?

Example 1:

1234×2

The method is simple. We pick the first digit of the left and multiply it by 2:

$1234 \times 2 = 2\ldots$

Then, we pick the second digit of the left and multiply it by 2:

$1234 \times 2 = 24\ldots$

And so on until the last digit:

$1234 \times 2 = 2468$

The result is 2468.

It was extremely simple. Surely you've wondered... What if you multiply the digit by 2 getting a two-digit number? In that case we must do a small adjustment.

Example 2:

392×2

We pick the first digit of the left, which is 3, and we divide it by 2.

$392 \times 2 = 6\ldots$

Now take the second digit and multiply it by 2. Then:

$392 \times 2 = 6(18)\ldots$ As noted, 18 is a two-digit number therefore 10 is subtracted from 18 and 1 is added to the previous digit of the partial result, which is 6.

$392 \times 2 = 78\ldots$ and continue the procedure.

$392 \times 2 = 784$

Therefore, the final result would be:

$392 \times 2 = 784$

Example 3:

$84725 \cdot 2$

$84725 \times 2 = 16\ldots$

$84725 \times 2 = 168\ldots$

$84725 \times 2 = 168(14)\ldots$

$84725 \times 2 = 1694\ldots$

$84725 \times 2 = 16944\ldots$

$84725 \times 2 = 16944(10)\ldots$

$84725 \times 2 = 169450$

Therefore:

$84725 \times 2 = 169450$

1.5.3 Back to the division by 5.

Now that we know how to multiply by 2, we are ready to perform any division by 5. We remind once again the rule for dividing by 5:

$$\frac{N}{5} = \frac{2N}{10} \tag{6}$$

Example:

$\frac{365}{5} = \ldots$ First multiply by 2 and then divide by 10.

$365 \times 2 = \ldots$

$365 \times 2 = 6\ldots$

$365 \times 2 = 6(12)\ldots$

$365 \times 2 = 72\ldots$

$365 \times 2 = 72(10)\ldots$

$365 \times 2 = 730$

Now divide by 10:

$\frac{730}{10} = 73$ Then,

$$\mathbf{\frac{365}{5} = 73}$$

Let's consider some other examples, now with decimal numbers.

Example 1:

$\frac{8.38}{5} = \ldots$

$8.38 \times 2 =$

$8.38 \times 2 = 16.\ldots$

$8.38 \times 2 = 16.6\ldots$

$8.38 \times 2 = 16.6(16)\ldots$

$8.38 \times 2 = 16.76$

$\frac{16.76}{10} = 1.676$ Then,

$$\mathbf{\frac{8.38}{5} = 1.676}$$

Example 2:

$\frac{324.6}{5} = (324.6 \times 2) : 10 = 649.2 : 10 = 64.92$

Example 3:

$\frac{0.0032}{5} = (0.0032 \times 2) : 10 = 0.0064 : 10 = 0.00064$

In these examples we just saw, the first thing we do is multiply by the number 2, then we move the decimal point one place to the left of the number previously obtained (which is equivalent to divide by 10 the number obtained in the first step).

Try the following exercises.

Divide the following numbers by 5. Apply the previous algorithm:

a) 423
b) 121432
c) 621240
d) 2.8
e) 32.1
f) 324.8
g) 143.22
h) 2314.31
i) 2233.44

1.5.4 How to divide by 25?

Dividing by 25 is equivalent to multiplying by 4 and then dividing that result by 100.

$$\frac{N}{25} = \frac{4N}{100} \tag{7}$$

Example:

$\frac{3211}{25} = (3211 \times 4) : 100 = 12844 : 100 = 128.44$

The result is 128.44.

Look at the following examples:

Example 1:

$\frac{7}{25} = \ldots$

25

$7 \times 4 = 28$ — The number is multiplied by 4.

$\frac{28}{100} = 0.28$ — When dividing by 100 move the decimal point two places to the left.

Example 2:

$\frac{275}{25} = \ldots$

$275 \times 4 = 1100$

$\frac{1100}{100} = 11.00$

The result is 11.

Example 3:

$\frac{8212}{25} = \ldots$

$8212 \times 4 = 32848$

$\frac{32848}{100} = 328.48$

The following example will be solved mentally, applying the rule of the division by 25 which we have just explained.

Divide $\frac{1111}{25} = \ldots$

4444 — Multiply the number by 4.

44.44 — Divide the number by 100.

Now we will do division by 25, in this case with decimal numbers and of course, using the previous procedure.

Example 1:

$\frac{320.1}{25} = (320.1 \times 4) : 100 = 1280.4 : 100 = 12.804$

Example 2:

$\frac{910.21}{25} =$

$910.21 \times 4 = 3640.84$ — First multiply the number by 4.

$\frac{3640.84}{100} = 36.4084$ — In the division by 100, we move the decimal point two places to the left.

The result is 36.4084.

Example 3:

$\frac{0.11}{25} =$

$0.11 \times 4 = 0.44$

$\frac{0.44}{100} = 0.0044$

Now let's solve it directly using the algorithm.

Example 4:

$\frac{1240.2}{25} =$

$= 4960.8$ — Multiplied by 4.

$= 49.608$ — Divided by 100.

Example 5:

$\frac{912.12}{25} = \frac{3648.48}{100} = 36.4848$

With practice you will be able to solve these calculations mentally, without any difficulty.

You are ready to solve the following exercises applying the rule for dividing by 25.

Divide the following numbers by 25:

a) 14
b) 159
c) 4215
d) 4121
e) 38122
f) 34.13
g) 346.2
h) 111.222
i) 201.202

1.5.5 How to divide by 12.5?

If we divide a number by 12.5, we multiply it by 8 and then divide the result by 100.

$$12.5N = \frac{8N}{100} \qquad (8)$$

$$\frac{3000}{12.5} = (3000 \times 8) : 100 = 24000 : 100 = 240$$

Consider some other examples.

Example 1:

$\frac{14}{12.5} = \ldots$

$= 14 \times 8 = 112$ — Multiply by 8.

$= \frac{112}{100} = 1.12$ — When dividing by 100 we move the decimal point two places to the left.

Example 2:

$\frac{210}{12.5} = (210 \times 8) : 100 = 1680 : 100 = 16.80$

The result is 16.8.

Example 3:

$\frac{9110}{12.5} = (9110 \times 8) : 100 = 72880 : 100 = 728.80$

The result is 728.8.

These exercises can be solved directly if we apply the algorithm.

Example 4:

Divide $\frac{110}{12.5} = \ldots$

880 — Multiplied by 8.

8.8 — Divided by 100.

Example 5:

$\frac{211}{12.5} = \ldots$

1688 — Multiplied by 8.

16.88 — Divided by 100.

Now we are going to divide decimal numbers by 12.5

Example 1:

$\frac{10.02}{12.5} = \ldots$

$10.02 \times 8 = 80.16$ — Multiply the number by 8.

$80.16 : 100 = 0.8016$ — Move the decimal point two places to the left.

The result is 0.8016.

Example 2:

$\frac{121.1}{12.5} = (121.1 \times 8) : 100 = 968.8 : 100 = 9.688$

The result is 9.688.

Example 3:

$\frac{0.31}{12.5} = (0.31 \times 8) : 100 = 2.48 : 100 = 0.0248$

The result is 0.0248.

Solve mentally, always applying the algorithm:

$\frac{20.01}{12.5} = \ldots$

160.08 — Multiplied by 8.

1.6008 — Divided by 100.

Try the following exercises.

Divide the following numbers by 12.5:

a) 11
b) 34
c) 510
d) 4000
e) 4211
f) 12.1
g) 21.1
h) 300.11

2 Complex multiplications.

In 2006 I participated in a Mental Calculation World Championship for the first time. It was held in Giessen, Germany. Since then I have had the opportunity to discuss many of the algorithms presented in this chapter with professional mental calculators. The following algorithms are considered the best for large multiplication problems.

2.1 Multiplying 2-digit numbers.

We start by knowing how to multiply two 2-digit numbers. The answer is obtained from right to left in only three simple steps:

Example 1:

$$\begin{array}{r} 76 \\ \times\ 42 \\ \hline 3192 \end{array}$$

Step 1: Multiply the ones digits ($6\times2 = 12$). Write the 2 as the partial answer and carry 1 to the next step.

Step 2: Cross multiply the tens digit of a number with the ones digit of the other number and add the previous carry ($7\times2 + 1 + 6\times4 = 39$). Write the 9 to the left of the partial result.

Step 3: Multiply the tens digits and add the previous carry ($7\times4 + 3 = 31$). We write 31 to the left of the partial result and you get the final result 3192.

Example 2:

$$\begin{array}{l} 38 \ - \ \text{Step 1: } 8\times5 = 40 \\ \times\ 65 \ - \ \text{Step 2: } 3\times5+4+8\times6 = 67 \\ \hline 2470 \ - \ \text{Step 3: } 3\times6+6 = 24 \end{array}$$

Now try practicing a few on your own:

a) $72 \times 16 =$
b) $68 \times 53 =$
c) $94 \times 49 =$
d) $61 \times 86 =$
e) $98 \times 97 =$

2.2 Multiplying 3-digit numbers.

For multiplying two 3-digit numbers just apply 5 simple steps:

$$
\begin{array}{r}
629 \\
\times\ 508 \\
\hline
319532
\end{array}
$$

Step 1: $9{\times}8 = 72$
Step 2: $2{\times}8 + 7 + 9{\times}0 = 23$
Step 3: $6{\times}8 + 2 + 2{\times}0 + 9{\times}5 = 95$
Step 4: $6{\times}0 + 9 + 2{\times}5 = 19$
Step 5: $6{\times}5 + 1 = 31$

Try the following on your own:

a) $718 \times 426 =$
b) $395 \times 174 =$
c) $686 \times 751 =$
d) $256 \times 794 =$
e) $975 \times 928 =$

2.3 Multiplying 4-digit numbers.

For multiplying two 4-digit numbers just apply 7 simple steps:

$$
\begin{array}{r}
3571 \\
\times\ 2384 \\
\hline
8513264
\end{array}
$$

Step 1: $1{\times}4 = 4$
Step 2: $7{\times}4 + 1{\times}8 = 36$
Step 3: $5{\times}4 + 3 + 7{\times}8 + 1{\times}3 = 82$
Step 4: $3{\times}4 + 8 + 5{\times}8 + 7{\times}3 + 1{\times}2 = 83$
Step 5: $3{\times}8 + 8 + 5{\times}3 + 7{\times}2 = 61$
Step 6: $3{\times}3 + 6 + 5{\times}2 = 25$
Step 7: $3{\times}2 + 2 = 8$

Try the following on your own:

a) $6305 \times 2012 =$
b) $1784 \times 3529 =$
c) $5912 \times 6132 =$
d) $4473 \times 9087 =$
e) $7613 \times 4808 =$
f) $9927 \times 9241 =$

2.4 Multiplying 8-digit numbers.

The following type of problem is often found in competition at the Mental Calculation World Championships:

$$49347632$$
$$\times\ 20581874$$

The competitors are only allowed to write the final answer on paper without writing any partial result. Since we are talking about multiplying two numbers of 8 digits each, many people wonder... But how can they do it?

The answer, however difficult it may seem, is SIMPLE. Through the method of "Cross Multiplication", the response will be calculated at each step. Let's describe the process:

49347632
× 20581874
_____
8 (multiply $2 \times 4 = 8$ and write 8 in the last digit)

49347632
× 20581874
_____
68 ($3 \times 4 + 7 \times 2 = 26$, write the 6 and carry 2)

49347632
× 20581874
_____
368 ($6 \times 4 + 3 \times 7 + 8 \times 2 + 2$ of previous carry $= 63$, write 3 and carry 6)

49347632
× 20581874
_____
2368 ($7 \times 4 + 6 \times 7 + 8 \times 3 + 1 \times 2 + 6 = 102$, write 2 and carry 10)

49347632
× 20581874
_____
22368 ($4 \times 4 + 7 \times 7 + 6 \times 8 + 1 \times 3 + 8 \times 2 + 10 = 142$, write 2 and carry 14)

49347632
× 20581874
_____
022368 ($3 \times 4 + 4 \times 7 + 7 \times 8 + 1 \times 6 + 8 \times 3 + 5 \times 2 + 14 = 150$)

49347632
× 20581874
_____
4022368 ($9 \times 4 + 3 \times 7 + 4 \times 8 + 7 \times 1 + 8 \times 6 + 5 \times 3 + 0 \times 2 + 15 = 174$)

49347632
× 20581874
_____
44022368 ($4 \times 4 + 9 \times 7 + 3 \times 8 + 4 \times 1 + 8 \times 7 + 5 \times 6 + 0 \times 3 + 2 \times 2 + 17 = 214$)

$$49347632$$
$$\times\ 20581874$$
$$\overline{744022368}\ (4 \times 7 + 9 \times 8 + 3 \times 1 + 4 \times 8 + 5 \times 7 + 0 \times 6 + 2 \times 3 + 21 = 197)$$

$$49347632$$
$$\times\ 20581874$$
$$\overline{6744022368}\ (4 \times 8 + 9 \times 1 + 3 \times 8 + 5 \times 4 + 0 \times 7 + 2 \times 6 + 19 = 116)$$

$$49347632$$
$$\times\ 20581874$$
$$\overline{66744022368}\ (4 \times 1 + 9 \times 8 + 3 \times 5 + 0 \times 4 + 2 \times 7 + 11 = 116)$$

$$49347632$$
$$\times\ 20581874$$
$$\overline{666744022368}\ (4 \times 8 + 9 \times 5 + 0 \times 3 + 2 \times 4 + 11 = 96)$$

$$49347632$$
$$\times\ 20581874$$
$$\overline{5666744022368}\ (4 \times 5 + 9 \times 0 + 2 \times 3 + 9 = 35)$$

$$49347632$$
$$\times\ 20581874$$
$$\overline{15666744022368}\ (4 \times 0 + 2 \times 9 + 3 = 21)$$

$$49347632$$
$$\times\ 20581874$$
$$\overline{1015666744022368}\ (4 \times 2 + 2 = 10)$$

Then 1015666744022368 will be the final answer.

The great thing about these algorithms is that it allows you to compute the answer by simply multiplying and adding small numbers. For beginners, it is best to start with multiplications of 2 by 2 and 3 by 3. Once you feel confident in your skills, then advance to larger numbers.

3 How to divide easier and faster.

Most people find division difficult. This is a problem that stems from the very first time students are taught how to do division in school. They were never really taught correctly in the first place. In addition to that, as this mathematical operation becomes more complex (when the results are not exact, i.e., with decimal places) the difficulty increases, because the calculation becomes more laborious, especially when the periods of the quotients are large.

The goal is to create an algorithm that allows us to divide easier and faster, while having the precision needed.

If you memorize a series of digits, then is possible to divide at an incredible speed.

In general, if we divide a number by **"n"** the result will have at most **n−1** length in the period (Ex: $\frac{45}{7}$ = 6.428571 period $\overline{428571}$, which has length 6). This is because at some point some other intermediate remainder in the division will repeat and thereafter the results are repeated over and over again. Remember that the only possible values of the remainders are: 1,2,3, ..., n-1, if remainder is 0 the division ends immediately and there is no period because we obtain the exact result. There are some numbers that when dividing by them do not necessarily give periods of length n−1, for example:

If we divide a number by 3 the length of the period is 1 (since it repeats the same digit of the period).

$\frac{14}{3}$ = 4.6 period $\overline{6}$ (period of length one).

If we divide a number by 11 the length of the period is 2 (since it repeated the same two digits of the period).

$\frac{226}{11}$ = 20.54 period $\overline{54}$ (period of length two).

However other numbers like 17 and 23 where the fractions not always represent a whole integer is found to have periods of length 16 and 22 respectively.

For example:

Suppose we want to divide by 17. If we divide $\frac{1}{17}$ = 0.0588235294117647058823529411764 7 ... or what is the same 0.0588235294117647 period $\overline{0588235294117647}$, then if we memorize that string of length 16 (0588235294117647), we can quickly know the result.

Let's demonstrate with the following example:

Suppose someone asks us how much is 61 divided by 17

$\frac{61}{17}$ = 3. ... 61 divided by 17 gives 3 and remainder 10 = 61 − 17×3, then add a 0 (note: do the division on paper so you can see the process with my explanation), we divide $\frac{100}{17}$ and get 5 with remainder 15 (3.5 ... is the partial result), then $\frac{150}{17}$ = 8 with remainder 14 (if remainder is equal to 10 or 15 then the sequence would repeat, but we said earlier that the fractions with denominator 17 have always period of length 16). Even at this stage we can get the whole answer because of the information we have of the partial result.

$$\frac{61}{17} = 3.58\ldots$$

How do we do it? It is simple, we search for the digits that appear after the decimal point of the partial answer, 58 in this case, in the sequence 0588235294117647 and we notice that appears in the 2nd and 3rd digit, then we can obtain the final result which is:

$$\frac{61}{17} = 3.5882352941176470 \text{ period } \overline{5882352941176470} \text{ (note that the period is circular).}$$

Then, if the period of 17 is memorized we only need to find two numbers after the decimal point (instead of 16 digits or perhaps more to realize of the repetition of the digits) to get the final result.

In the case of 23, the sequence is 0434782608695652173913 then if we divide:

$$\frac{518}{23} = 22.52\ldots \text{ We will know instantly that the full answer will be:}$$

$$\frac{518}{23} = 22.5217391304347826086956 \text{ period } \overline{5217391304347826086956}.$$

Obviously, having to memorize every sequence is not feasible, but memorizing all the sequences of dividends less than 50, using the table that we offer bellow, would make division the easiest operation of all, instead of the hardest.

Divisor	Period
1,2,4,5,8,10,16,20,25,32,40,50	Empty (no period)
3,6,12,15,24,30,48	$\overline{3}, \overline{6}$
7,14,28,35	$\overline{142857}$
9,18,36,45	$\overline{1}, \overline{2}, \overline{3}, \overline{4}, \overline{5}, \overline{6}, \overline{7}, \overline{8}$
11,22,44	$\overline{09}, \overline{18}, \overline{27}, \overline{36}, \overline{45}$
13,26	$\overline{076923}, \overline{153846}$
17,34	$\overline{0588235294117647}$
19,38	$\overline{052631578947368421}$
21,42	$\overline{047619}, \overline{095238}, \overline{142857}, \overline{3}, \overline{6}$
23,46	$\overline{0434782608695652173913}$
27	$\overline{037}, \overline{074}, \overline{1}, \overline{148}, \overline{185}, \overline{2}, \overline{259}, \overline{296}, \overline{3}, \overline{4}, \overline{5}, \overline{6}, \overline{7}, \overline{8}$
29	$\overline{0344827586206896551724137931}$
31	$\overline{032258064516129}, \overline{096774193548387}$
33	$\overline{03}, \overline{06}, \overline{09}, \overline{12}, \overline{15}, \overline{18}, \overline{24}, \overline{27}, \overline{3}, \overline{36}, \overline{39}, \overline{45}, \overline{48}, \overline{57}, \overline{6}, \overline{69}, \overline{78}$
37	$\overline{027}, \overline{054}, \overline{081}, \overline{135}, \overline{162}, \overline{189}, \overline{243}, \overline{297}, \overline{378}, \overline{459}, \overline{486}, \overline{567}$
39	$\overline{025641}, \overline{051282}, \overline{076923}, \overline{153846}, \overline{179487}, \overline{3}, \overline{358974}, \overline{6}$
41	$\overline{02439}, \overline{04878}, \overline{07317}, \overline{09756}, \overline{12195}, \overline{14634}, \overline{26829}, \overline{36585}$
43	$\overline{023255813953488372093}, \overline{046511627906976744186}$
47	$\overline{0212765957446808510638297872340425531914893617}$
49	$\overline{020408163265306122448979591836734693877551}, \overline{142857}$

Table of periods up to 50.

4 Calculating square roots of a six-digit number.

Competitions of fast mental calculation are held, as is the Mental Calculation World Championship which establishes rules for the competitions. For instance, you must solve ten exercises of square roots in a time of ten minutes, so you have to solve each problem in at most one minute.

We are going to calculate square roots of numbers of six digits, with an accuracy of five digits after the decimal point.

If you use a conventional method will not be easy, but with other efficient methods, this can be done.

One method is the Duplex Numeric System Method (number D), which is as follows:

To obtain the duplex number, the procedure consists into multiply the first digit by the last digit, the second digit by the second last and so on. For an even number of figures all the products are added and the result multiplied by 2. For an odd number of figures all the products are added (except the middle digit), the total is multiplied by 2 and added the square of the middle digit.

Let's describe the procedure with an example and you will see that it is not impossible to do.

Examples:

Number	Number D (Duplex)
5	$5 \times 5 = 25$
34	$(3 \times 4) \times 2 = 24$
428	$(4 \times 8) \times 2 + 2 \times 2 = 68$
354	$(3 \times 4) \times 2 + 5 \times 5 = 49$
7495	$(7 \times 5 + 4 \times 9) \times 2 = 71 \times 2 = 142$
4321	$(4 \times 1 + 3 \times 2) \times 2 = 20$
2356	$(2 \times 6 + 3 \times 5) \times 2 = 27 \times 2 = 54$
26937	$(2 \times 7 + 6 \times 3) \times 2 + 9 \times 9 = 32 \times 2 + 81 = 145$
13542	$(1 \times 2 + 3 \times 4) \times 2 + 5 \times 5 = 14 \times 2 + 25 = 53$
231426	$(2 \times 6 + 3 \times 2 + 1 \times 4) \times 2 = 22 \times 2 = 44$
1234321	$(1 \times 1 + 2 \times 2 + 3 \times 3) \times 2 + 4 \times 4 = 14 \times 2 + 16 = 44$

After understanding the definition of the number Duplex and have analyzed the examples above, we are going to calculate the square root of 530179.

Starting with the ones digit, you form groups every two digits. As we form three groups, we know that the integer value of the root will be a 3-digit number.

Consider the example:
$\sqrt{530179}$

53 01 79 — We determine the nearest integer below to the root of 53. In this case is 7, because $7 \times 7 = 49$

7...

53 01 79 — Put the number 7 above the number 53.

— The number 7 is multiplied by 2, ie $7 \times 2 = 14$
(The number 14 will be considered for all the calculations in this exercise our DIVISOR)

7...
53 01 79

$53 - 49 = 4$ — We subtract the number 49, because that is the square of the number 7.
40 — This number is formed by the previous remainder and the number 0 of the next group.

$\frac{40}{14} = 2$ and remainder 12.

72...
53 01 79 — Now put the number 2, forming the 72 ... as a PARTIAL ANSWER.

121 — This number is formed from the remainder of the division by 14 (in this case 12) and the second digit of the second group.

Here we applied what we learned about the Duplex number. We calculate the Duplex to the partial answer, except for the first number (the procedure will always be the same). Duplex of 2 is $2 \times 2 = 4$

$121 - 4 = 117$ — Subtract the duplex number to 121 (4 in this case).

$\frac{117}{14} = 8$ and remainder 5.

Include the quotient of the division in my PARTIAL ANSWER, in my example is formed the number 728.

728...
53 01 79

$57 - 32 = 25$ — The number 57 is formed from the remainder of the division of $\frac{117}{14}$ and the number 7 which is the first digit of the third group.

Calculate the Duplex of 28, would be $(2 \times 8) \times 2 = 32$

As mentioned initially, my result is a number with three digits before the decimal point, then it is time to we write the decimal point.

$\frac{25}{14} = 1$ and remainder 11.

Include the quotient of the division in the PARTIAL ANSWER. In the example we have so far as partial answer the number 728.1.

728.1...

53 01 79

119 — This number is formed by the remainder of the division by 14 and the number 9 which is the second digit of the third group.

119 – 68 = 51 — Calculate the Duplex of 281 = $(2 \times 1) \times 2 + 8 \times 8 = 4 + 64 = 68$.

$\frac{51}{14} = 3$ and remainder 9.
The quotient 3 is placed in the PARTIAL ANSWER.

728.13...
53 01 79

We said that the square root would be calculated with an accuracy of five places after the decimal point. From now to form the number to which the duplex is subtracted we place a 0 next to it.

90

90 – 28 = 62 — Calculate the Duplex of 2813 = $(2 \times 3 + 8 \times 1) \times 2 = (6 + 8) \times 2 = 28$

$\frac{62}{14} = 4$ and remainder 6.
The quotient 4 is placed in the PARTIAL ANSWER.

60 – 65 = no positive value — Calculate the Duplex of 28134 = $(2 \times 4 + 8 \times 3) \times 2 + 1 \times 1 = 65$

Since 60 is less than 65 we can not do the subtraction (in the set of natural numbers). When this happens to us, we reduce the previous quotient to 1, in our example the quotient of $\frac{62}{14}$ which is 4 is reduced to 3 and we increase the remainder, obtaining this:

$\frac{62}{14} = 3$ and remainder 20.

Change the last digit of my PARTIAL ANSWER. In this case it was 4 but we needed to reduce the quotient in 1, getting 3.

728.133...
53 01 79

200 — Add a 0 to the remainder.
200 – 61 = 139 — Calculate the Duplex of 28133 = $(2 \times 3 + 8 \times 3) \times 2 + 1 \times 1 = (6 + 24) \times 2 + 1 = 61$

$\frac{139}{14} = 9$ and remainder 13.
The quotient 9 is placed in the PARTIAL ANSWER.

728.1339...
53 01 79

130 — Add a zero to the remainder.

$130 - 90 = 40$ — Calculate the Duplex of $281339 = (2 \times 9 + 8 \times 3 + 1 \times 3) \times 2 = (18 + 24 + 3) \times 2 = 90$

$\frac{40}{14} = 2$ and remainder 12.

The quotient 2 is placed in the PARTIAL ANSWER.

120 — Add a zero to the remainder.

$120 - 167 =$ no positive value — Calculate the Duplex of $2813392 = (2 \times 2 + 8 \times 9 + 1 \times 3) \times 2 + 3 \times 3 = (4 + 72 + 3) \times 2 + 9 = 167$

As 120 is less than 167 then the previous quotient is decreased by 1.

$\frac{40}{14} = 1$ and remainder 26. — The last digit of my answer is reduced by 1.

728.13391...
53 01 79

$260 - 163 = 97$ — Calculate the Duplex of $2813391 = (2 \times 1 + 8 \times 9 + 1 \times 3) \times 2 + 3 \times 3 = (2 + 72 + 3) \times 2 + 9 = 163$

Then, 97 can be divided by 14 so that finally the partial answer until the eighth figure is 728.13391. (This partial response is not rounded).

5 The perpetual calendar.

5.1 Brief History:

It was the great Emperor Julius Caesar who replaced the Egyptian calendar, which was based on exactly 365 days. He replaced it with a new calendar called the Julian calendar, which had an average year of 365,25 days and contained a leap year of 366 days every four years to adjust the length of a year. However, later calculations showed that the period of rotation of the Earth to be about 365.2422 days.

It was not until 1582 when Pope Gregory considered that leap years would be exactly the multiples of 4, except those that are divisible by 100 and not divisible by 400. With this arrangement, the modern calendar has an annual average of 365.2425 days, which is closer to the true value.

5.2 Fundamental lemma in obtaining the algorithm.

In this section we will propose a lemma that will help us to develop the algorithm:

Lemma: Every four hundred years the days of the week coincide.

Notice how interesting is it. Imagine if we know the day it was published the novel "Don Quixote" by Cervantes, we can know what day of the week fell just by checking a simple calendar of 2005, because it was published exactly 400 years earlier in 1605.

Applying the previous lemma we only need to create an algorithm valid for an interval of four hundred years.

5.3 Values to be memorized.

The algorithm showed below is simple. We only need to memorize a few values.

Month	Value
January	5
February	1
March	1
April	4
May	6
June	2
July	4
August	0
September	3
October	5
November	1
December	3

Table 1. Table of values associated with each month.

For each century we also have to memorize some values:

Century	Value
1700–1799	5
1800–1899	3
1900–1999	1
2000–2099	0

Table 2. Table of values associated with each century.

5.4   The algorithm.

Input: f (d/m/y).

Given:

Sum = 0. // Initialize the current sum.

Year = [y mod 100] (remainder of the division of **y** by 100).

Century = $[\frac{y}{100}]$

T1: values of table 1. (January = 5, February = 1,..., December = 3).

T2: values of table 2. // Index start at 0.

E: enumerator of the days of the week (Sunday=0, Monday=1, Tuesday=2, Wednesday=3, Thursday=4, Friday=5, Saturday=6).

Answer = ?

Sum = Year + $[\frac{Year}{4}]$ + 1

If Year mod 4 = 0 and m = 2

Sum = Sum – 1 // Here I remove the 29/02/2000+Year, which has not elapsed.

Sum = Sum + d + T1[m] + T2[Century].

If Year = 0 and Century mod 4 = 0 and m = 2

Sum = Sum + 1

Answer = E^{-1}(Sum mod 7)

Example 1:

January 28, 1853:

Input: f (28/1/1853)

Sum = 53 + $[\frac{53}{4}]$ + 1 + 28 + T1 [January] + T2 [1800–1899].

Sum = 53 + 13 + 1 + 28 + 5 + 3 = 103.

When dividing 103 by 7 the remainder of the division is 5 (103 mod 7 = 5).

Answer = E^{-1}(5). If you get 5 is because the answer is **Friday**.

The Apostle Jose Marti, Cuba's National Hero, was born on a **Friday**.

Example 2:

July 14, 1789:
Input: f (14/7/1789)
Sum = $89 + [\frac{89}{4}] + 1 + 14 +$ T1 [July] + T2 [1700–1799].
Sum = $89 + 22 + 1 + 14 + 4 + 5 = 135$.

When dividing 135 by 7 the remainder of the division is 2 (135 mod 7 = 2).
Answer = $E^{-1}(2)$. If you get 2 is because the answer is **Tuesday**.

The storming of the Bastille occurred on a **Tuesday**.

Example 3:

April 26, 2082:
Input: f (26/4/2082)
Sum = $82 + [\frac{82}{4}] + 1 + 26 +$ T1 [April] + T2 [2000–2099].
Sum = $82 + 20 + 1 + 26 + 4 + 0 = 133$.

When dividing 133 by 7 the remainder of the division is 0 (133 mod 7 = 0).
Answer = $E^{-1}(0)$. If you get 0 is because the answer is **Sunday**.

My centenary will be on a **Sunday**.

This algorithm out-performs other existing algorithms because of its low number of arithmetic operations. Personally, using this method I have managed to become the World Record Holder in the category. In competition, I calculated 93 dates in 1 minute.

I sincerely hope with all my heart that this book has motivated you to seriously consider that mental arithmetic is a powerful tool. With practice you can improve your mental agility about sevenfold. I wish you good luck in the beautiful numeric journey that you have just begun.

6 Answers to exercises.

Section 1.3.3

a) $825 \times 5 = \frac{825}{2} \times 10 = 412.5 \times 10 = 4125$

b) $42362 \times 5 = \frac{42362}{2} \times 10 = 21181 \times 10 = 211810$

c) $3242446 \times 5 = \frac{3242446}{2} \times 10 = 1621223 \times 10 = 16212230$

d) $102 \times 5 = \frac{102}{2} \times 10 = 51 \times 10 = 510$

e) $1004 \times 5 = \frac{1004}{2} \times 10 = 502 \times 10 = 5020$

f) $305.4 \times 5 = \frac{305.4}{2} \times 10 = 152.7 \times 10 = 1527$

g) $1084.24 \times 5 = \frac{1084.24}{2} \times 10 = 542.12 \times 10 = 5421.2$

h) $0.42 \times 5 = \frac{0.42}{2} \times 10 = 0.21 \times 10 = 2.1$

i) $14.87 \times 5 = \frac{14.87}{2} \times 10 = 7.435 \times 10 = 74.35$

j) $14.846 \times 5 = \frac{14.846}{2} \times 10 = 7.423 \times 10 = 74.23$

Section 1.3.4

a) $16 \times 25 = \frac{16}{4} \times 100 = 4 \times 100 = 400$

b) $52 \times 25 = \frac{52}{4} \times 100 = 13 \times 100 = 1300$

c) $124 \times 25 = \frac{124}{4} \times 100 = 31 \times 100 = 3100$

d) $400 \times 25 = \frac{400}{4} \times 100 = 100 \times 100 = 10000$

e) $24.48 \times 25 = \frac{24.48}{4} \times 100 = 6.12 \times 100 = 612$

f) $12.24 \times 25 = \frac{12.24}{4} \times 100 = 3.06 \times 100 = 306$

Section 1.3.5

a) $24 \times 12.5 = \frac{24}{8} \times 100 = 3 \times 100 = 300$

b) $48 \times 12.5 = \frac{48}{8} \times 100 = 6 \times 100 = 600$

c) $32 \times 12.5 = \frac{32}{8} \times 100 = 4 \times 100 = 400$

d) $72 \times 12.5 = \frac{72}{8} \times 100 = 9 \times 100 = 900$

e) $3248 \times 12.5 = \frac{3248}{8} \times 100 = 406 \times 100 = 40600$

f) $7288 \times 12.5 = \frac{7288}{8} \times 100 = 911 \times 100 = 91100$

g) $16.96 \times 12.5 = \frac{16.96}{8} \times 100 = 2.12 \times 100 = 212$

h) $248.4 \times 12.5 = \frac{248.4}{8} \times 100 = 31.05 \times 100 = 3105$

Section 1.4.1

a) $26 \times 11 = 2(2+6)6 = 286$

b) $45 \times 11 = 4(4+5)5 = 495$

c) $83 \times 11 = 8(8+3)3 = 913$

d) $64 \times 11 = 6(6+4)4 = 704$

e) $78 \times 11 = 7(7+8)8 = 858$

f) $96 \times 11 = 9(9+6)6 = 1056$

g) $429 \times 11 = \ldots 9 = \ldots(2+9)9 = \ldots 19 = \ldots(4+2+1)19 = \ldots 719 = 4719$

h) $62937 \times 11 = 6(6+2+1)(2+9+1)(9+3+1)(3+7)7 = 692307$

i) $84703 \times 11 = (8+1)(8+4+1)(4+7)(7+0)(0+3)3 = 931733$

j) $3904168265 \times 11 = (3+1)(3+9)(9+0)(0+4)(4+1)(1+6+1)(6+8+1)(8+2)(2+6+1)(6+5)5 = 42945850915$

Section 1.4.2

a) $718 \times 22 = [7 \times 2 + 1][(7+1) \times 2 + 1][(1+8) \times 2 + 1][8 \times 2] = 15796$

b) $6603 \times 55 = [6 \times 5 + 6][(6+6) \times 5 + 3][(6+0) \times 5 + 1][(0+3) \times 5 + 1][3 \times 5] = 363165$

c) $38247 \times 77 = [3 \times 7 + 8][(3+8) \times 7 + 7][(8+2) \times 7 + 5][(2+4) \times 7 + 8][(4+7) \times 7 + 4][7 \times 7] = 2945019$

d) $5082643 \times 99 = [5 \times 9 + 5][(5+0) \times 9 + 8][(0+8) \times 9 + 9][(8+2) \times 9 + 8][(2+6) \times 9 + 9][(6+4) \times 9 + 6][(4+3) \times 9 + 2][3 \times 9]$
$= 503181657$

Section 1.4.3

a) $18 \times 45 = 18 \times (50 - 5) = 900 - 90 = 810$

b) $21 \times 54 = 21 \times (60 - 6) = 1260 - 126 = 1134$

c) $15 \times 63 = 15 \times (70 - 7) = 1050 - 105 = 945$

d) $34 \times 27 = 34 \times (30 - 3) = 1020 - 102 = 918$

e) $92 \times 36 = 92 \times (40 - 4) = 3680 - 368 = 3312$

f) $43 \times 45 = 43 \times (50 - 5) = 2150 - 215 = 1935$

g) $14 \times 54 = 14 \times (60 - 6) = 840 - 84 = 756$

h) $23 \times 63 = 23 \times (70 - 7) = 1610 - 161 = 1449$

i) $54 \times 72 = 54 \times (80 - 8) = 4320 - 432 = 3888$

j) $35 \times 81 = 35 \times (90 - 9) = 3150 - 315 = 2835$

k) $320 \times 27 = 320 \times (30 - 3) = 9600 - 960 = 8640$

l) $121 \times 45 = 121 \times (50 - 5) = 6050 - 605 = 5445$

m) $184 \times 45 = 184 \times (50 - 5) = 9200 - 920 = 8280$

n) $143 \times 36 = 143 \times (40 - 4) = 5720 - 572 = 5148$

1) $385 \times 5 = \frac{385}{2} \times 10 = 192.5 \times 10 = 1925$

2) $436 \times 5 = \frac{436}{2} \times 10 = 218 \times 10 = 2180$

3) $38.46 \times 5 = \frac{38.46}{2} \times 10 = 19.23 \times 10 = 192{,}3$

4) $602.8 \times 5 = \frac{602.8}{2} \times 10 = 301.4 \times 10 = 3014$

5) $488.4 \times 5 = \frac{488.4}{2} \times 10 = 244.2 \times 10 = 2442$

6) $2084.64 \times 5 = \frac{2084.64}{2} \times 10 = 1042.32 \times 10 = 10423.2$

7) $720 \times 25 = \frac{720}{4} \times 100 = 180 \times 100 = 18000$

8) $64 \times 25 = \frac{64}{4} \times 100 = 16 \times 100 = 1600$

9) $16.88 \times 25 = \frac{16.88}{4} \times 100 = 4.22 \times 100 = 422$

10) $32 \times 12{,}5 = \frac{32}{8} \times 100 = 4 \times 100 = 400$

11) $16 \times 12.5 = \frac{16}{8} \times 100 = 2 \times 100 = 200$

12) $111 \times 27 = 111 \times (30 - 3) = 3330 - 333 = 2997$

13) $26 \times 36 = 26 \times (40 - 4) = 1040 - 104 = 936$

14) $44 \times 63 = 44 \times (70 - 7) = 3080 - 308 = 2772$

15) $14 \times 18 = 14 \times (20 - 2) = 280 - 28 = 252$

16) $143 \times 45 = 143 \times (50 - 5) = 7150 - 715 = 6435$

17) $26 \times 54 = 26 \times (60 - 6) = 1560 - 156 = 1404$

18) $32 \times 72 = 32 \times (80 - 8) = 2560 - 256 = 2304$

19) $48 \times 81 = 48 \times (90 - 9) = 4320 - 432 = 3888$

20) $184 \times 27 = 184 \times (30 - 3) = 5520 - 552 = 4968$

21) $143 \times 18 = 143 \times (20 - 2) = 2860 - 286 = 2574$

Section 1.5.3

a) $\frac{423}{5} = (423 \times 2) : 10 = 846 : 10 = 84.6$

b) $\frac{121432}{5} = (121432 \times 2) : 10 = 242864 : 10 = 24286.4$

c) $\frac{621240}{5} = (621240 \times 2) : 10 = 1242480 : 10 = 124248$

d) $\frac{2.8}{5} = (2.8 \times 2) : 10 = 5.6 : 10 = 0.56$

e) $\frac{32.1}{5} = (32.1 \times 2) : 10 = 64.2 : 10 = 6.42$

f) $\frac{324.8}{5} = (324.8 \times 2) : 10 = 649.6 : 10 = 64.96$

g) $\frac{143.22}{5} = (143.22 \times 2) : 10 = 286.44 : 10 = 28.644$

h) $\frac{2314.31}{5} = (2314.31 \times 2) : 10 = 4628.62 : 10 = 462.862$

i) $\frac{2233.44}{5} = (2233.44 \times 2) : 10 = 4466.88 : 10 = 446.688$

Section 1.5.4

a) $\frac{14}{25} = (14 \times 4) : 100 = 56 : 100 = 0.56$

b) $\frac{159}{25} = (159 \times 4) : 100 = 636 : 100 = 6.36$

c) $\frac{4215}{25} = (4215 \times 4) : 100 = 16860 : 100 = 168.6$

d) $\frac{4121}{25} = (4121 \times 4) : 100 = 16484 : 100 = 164.84$

e) $\frac{38122}{25} = (38122 \times 4) : 100 = 152488 : 100 = 1524.88$

f) $\frac{34.13}{25} = (34.13 \times 4) : 100 = 136.52 : 100 = 1.3652$

g) $\frac{346.2}{25} = (346.2 \times 4) : 100 = 1384.8 : 100 = 13.848$

h) $\frac{111.222}{25} = (111.222 \times 4) : 100 = 444.888 : 100 = 4.44888$

i) $\frac{201.202}{25} = (201.202 \times 4) : 100 = 804.808 : 100 = 8.04808$

Section 1.5.5

a) $\frac{11}{12.5} = (11 \times 8) : 100 = 88 : 100 = 0.88$

b) $\frac{34}{12.5} = (34 \times 8) : 100 = 272 : 100 = 2.72$

c) $\frac{510}{12.5} = (510 \times 8) : 100 = 4080 : 100 = 40.8$

d) $\frac{4000}{12.5} = (4000 \times 8) : 100 = 32000 : 100 = 320$

e) $\frac{4211}{12.5} = (4211 \times 8) : 100 = 33688 : 100 = 336.88$

f) $\frac{12.1}{12.5} = (12.1 \times 8) : 100 = 96.8 : 100 = 0.968$

g) $\frac{21.1}{12.5} = (21.1 \times 8) : 100 = 168.8 : 100 = 1.688$

h) $\frac{300.11}{12.5} = (300.11 \times 8) : 100 = 2400.88 : 100 = 24.0088$

Section 2.1

a) 72 — Step 1: $2 \times 6 = 12$
    $\times\ 16$ — Step 2: $7 \times 6 + 1 + 2 \times 1 = 45$
    1152 — Step 3: $7 \times 1 + 4 = 11$

b) $68 \times 53 = 3604$
c) $94 \times 49 = 4606$
d) $61 \times 86 = 5246$
e) $98 \times 97 = 9506$

Section 2.2

a)
$$\begin{array}{r} 718 \\ \times\ 426 \\ \hline 305868 \end{array}$$

Step 1: $8 \times 6 = 48$
Step 2: $1 \times 6 + 4 + 8 \times 2 = 26$
Step 3: $7 \times 6 + 2 + 1 \times 2 + 8 \times 4 = 78$
Step 4: $7 \times 2 + 7 + 1 \times 4 = 25$
Step 5: $7 \times 4 + 2 = 30$

b) $395 \times 174 = 68730$
c) $686 \times 751 = 515186$
d) $256 \times 794 = 203264$
e) $975 \times 928 = 904800$

Section 2.3

a)
$$\begin{array}{r} 6305 \\ \times\ 2012 \\ \hline 12685660 \end{array}$$

Step 1: $5 \times 2 = 10$
Step 2: $0 \times 2 + 1 + 5 \times 1 = 6$
Step 3: $3 \times 2 + 0 \times 1 + 5 \times 0 = 6$
Step 4: $6 \times 2 + 3 \times 1 + 0 \times 0 + 5 \times 2 = 25$
Step 5: $6 \times 1 + 2 + 3 \times 0 + 0 \times 2 = 8$
Step 6: $6 \times 0 + 3 \times 2 = 6$
Step 7: $6 \times 2 = 12$

b) $1784 \times 3529 = 6295736$
c) $5912 \times 6132 = 36252384$
d) $4473 \times 9087 = 40646151$
e) $7613 \times 4808 = 36603304$
f) $9927 \times 9241 = 91735407$

www.ingramcontent.com/pod-product-compliance
Lightning Source LLC
Chambersburg PA
CBHW051059180526
45172CB00002B/697